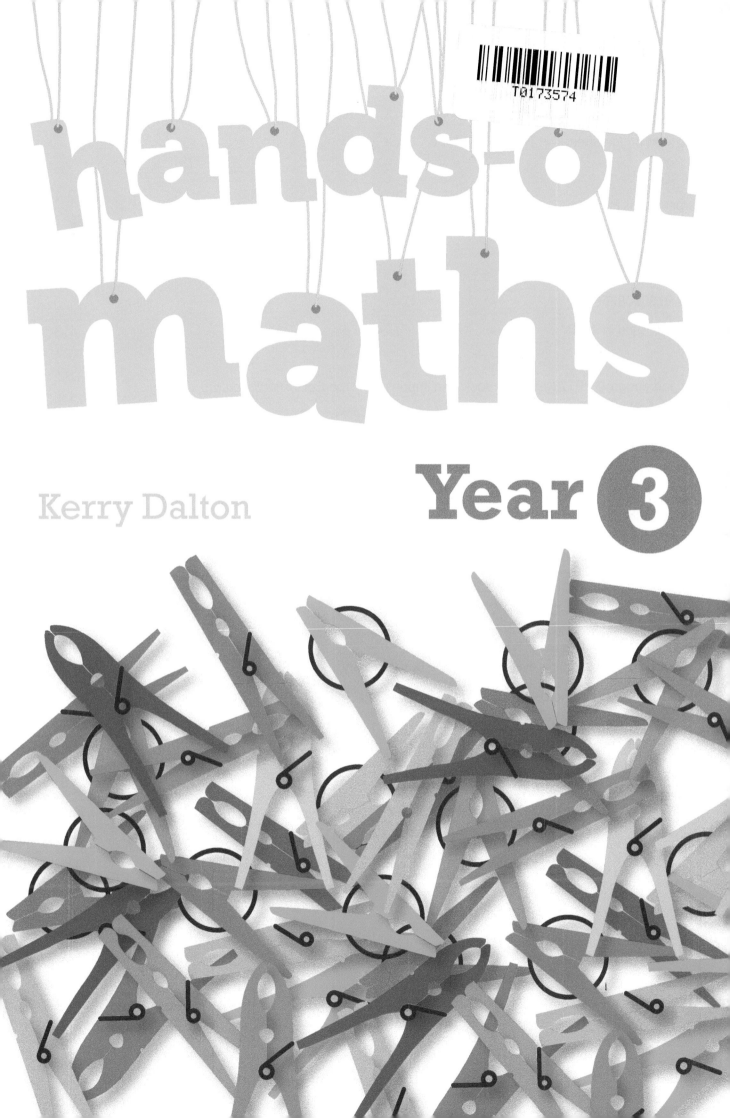

hands-on maths

Year 3

Kerry Dalton

Published by Keen Kite Books
An imprint of HarperCollins*Publishers* Ltd
The News Building
1 London Bridge Street
London
SE1 9GF

HarperCollins Publishers,
Macken House, 39/40 Mayor Street Upper,
Dublin 1,
D01 C9W8 Ireland

ISBN 9780008266974

First published in 2017
10 9

Text and design © 2017 Keen Kite Books, an imprint of HarperCollins*Publishers* Ltd
Author: Kerry Dalton

Series Concept and Commissioning: Shelley Teasdale and Michelle I'Anson
Project Manager: Fiona Watson
Editor: Denise Moulton
Cover Design: Anthony Godber
Text Design and Layout: Contentra Technologies
Production: Natalia Rebow
A CIP record of this book is available from the British Library.

Printed and bound in the UK by Ashford Colour Press Ltd

Contents

Year 3 aims and objectives

Hands-on Maths Year 3 encourages pupils to enjoy a range of mathematical concepts through a practical and hands-on approach. Using a range of everyday objects and common mathematical resources, pupils will explore and represent key mathematical concepts. These concepts are linked directly to the National Curriculum 2014 objectives for Year 3. Each objective will be investigated over the course of the week using a wide range of hands-on approaches such as Dienes, place-value counters, abacus models, playing cards, dice, place-value grids, practical problems and a mix of individual and paired work. The mathematical concepts are explored in a variety of contexts to give pupils a richer and deeper learning experience, which enables mastery to be attained.

Year 3 programme and overview of objectives

Topic	Week 1	Week 2	Week 3	Week 4	Week 5	Week 6
Counting	Count from 0 in multiples of 4	Count from 0 in multiples of 8	Count from 0 in multiples of 4 and 8 *(relational understanding)*	Count from 0 in multiples of 100	Count from 0 in multiples of 50	Count from 0 in multiples of 50 and 100 *(relational understanding)*
Place value	Recognise the place value of each digit in a three-digit number	Compare and order numbers up to 1000	Read and write numbers up to 1000 in numerals	Read and write numbers up to 1000 in words	Solve number problems using place-value ideas	Solve practical problems using place-value ideas
Representing numbers	Identify, represent and estimate numbers using different representations *(place value)*	Identify, represent and estimate numbers using different representations *(place value)*	Identify, represent and estimate numbers using different representations *(money)*	Identify, represent and estimate numbers using different representations *(number lines)*	Identify, represent and estimate numbers using different representations *(scaling)*	Identify, represent and estimate numbers using different representations *(partitioning)*
Addition and subtraction	Add mentally a three-digit number and ones	Subtract mentally a three-digit number and ones	Add mentally a three-digit number and tens	Subtract mentally a three-digit number and tens	Add mentally a three-digit number and hundreds	Subtract mentally a three-digit number and hundreds

Topic	Week 1	Week 2	Week 3	Week 4	Week 5	Week 6
Multiplication and division	Recall and use multiplication and division facts for the 3 multiplication table	Recall and use multiplication and division facts for the 4 multiplication table	Recall and use multiplication and division facts for the 8 multiplication table	Solve problems, including missing number problems, involving multiplication and division	Solve problems involving multiplication, including positive integer scaling problems	Solve problems involving multiplication *(including commutativity)*
Fractions	Count up and down in tenths	Recognise, find and write fractions of a discrete set of objects	Recognise and use fractions as numbers	Recognise, and show fractions using diagrams, equivalent fractions with small denominators	Add and subtract fractions with the same denominator within one whole	Compare and order unit fractions, and fractions with the same denominators

Introduction

The *Hands-on maths* series of books aims to develop the use of readily available manipulatives such as toy cars, shells and counters to support understanding in maths. The series supports a concrete–pictorial–abstract approach to help develop pupils' mastery of key National Curriculum objectives.

Each title covers six topic areas from the National Curriculum (counting; representing numbers; place value; the four number operations: addition and subtraction and multiplication and division; and fractions). Each area is covered during a six-week unit, with an easy-to-implement 10-minute activity provided for each day of the week. Photos are included for each activity to support delivery.

Hands-on maths enables a deep interrogation of the curriculum objectives, using a broad range of approaches and resources. It is not intended that schools purchase additional or specialist equipment to deliver the sessions; in fact, it is hoped that pupils will very much help to prepare resources for the different units, using a range of natural, formal and typical maths resources found in most classrooms and schools. This will help pupils to find ways to independently gain a deep understanding and enjoyment of maths.

A typical 'hands-on' classroom will have a good range of resources, both formal and informal. These may include counters, playing cards, coins, Dienes, dominoes, small objects such as toy cars and animals, Cuisenaire rods, 100 squares and hoops.

There is no requirement to use *only* the resources seen in the photographs that accompany each activity. Cubes may look like those in the green bowl, or will be just as effective if they look like the ones in the blue bowl. They serve the same purpose in helping pupils understand what the cubes represent.

Resources

Hands-on Maths uses a range of formal, informal and 'typical' resources found in most classrooms and schools. To complete the activities in this book, it is expected that teachers will have the following resources readily available:

- whiteboards and pens for individual pupils and pairs of pupils
- Dienes and Cuisenaire rods
- dice, coins and bead strings
- a range of cards, including playing cards, place-value arrow cards and digit cards

- collections of objects that pupils are interested in and want to count, such as toy cars, toy animals and shells
- bowls / containers to store sets of resources in, making it easy for pupils to handle and use the objects

- ten frames (these could be egg boxes, ice-cube trays, printed frames or something pupils have created themselves)

- number lines and 100 squares – lots of different types and styles: printed, home-made, interactive, digital or practical … whatever you prefer, and whatever is handy. (For 100 squares, there is, of course, the 1–100 or 0–99 choice to make; both work and it is best to choose whatever works for the class. Both offer a slight difference in place-value perspective, with 0–99 giving the 'zero as a place holder' emphasis, while the 1–100 version helps pupils to visualise the position of 100 in relation to the two-digit numbers.)

- counters and cubes – lots of them! Many of the activities require counters and cubes to be readily available. The cubes can be any size and any colour: what the cubes represent is the most important factor.

Assessment, cross-curricular links and vocabulary

Maths is a truly unique, creative and exciting discipline that can provide pupils with the opportunity to delve deeply into core concepts. Maths is found all around us, every day, in many different forms. It complements the principles of science, technology and engineering.

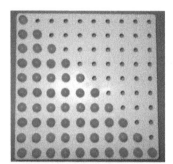

Hands-on maths provides ideas that can be adapted to suit the broad range of needs in our classrooms today. These ideas can be used as a starting point for assessment – before, during or after teaching a particular topic has taken place. The activities are intended to be flexible enough to be used with a whole class and can, of course, be differentiated to suit individual pupils in a class.

The activities can be adapted to link to other subject areas and interests. For example, a suggestion to use farm animals may link well to a science unit on classification or food chains; alternatively, the resource could be substituted with bugs if minibeasts is an area of interest for pupils. Teachers can be as flexible as they wish with the activities and resources – class teachers know their pupils best.

Spoken language is underpinned in maths by the unique mathematical vocabulary pupils need to be able to use fluently in order to demonstrate their reasoning skills and show mathematical proof. The correct, regular and secure use of mathematical language is key to pupils' understanding; it is the way in which they reason verbally, negotiate conceptual understanding and build secure foundations for a love of mathematics and all that it brings. Each unit in *Hands-on maths* identifies a range of vocabulary that is typical, but by no means limited to, that particular unit. The way the vocabulary is used and incorporated into activities is down to individual style and preference and, as with all of the resources in the book, will be very much dependent on the needs of each individual class. A blank template for creating vocabulary cards is included at the back of this book.

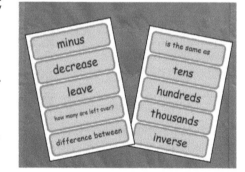

Week 1: Counting

Count from 0 in multiples of 4

Resources: cubes, 100 squares, number lines, sticky notes, playing cards

Vocabulary: number, zero, one, two, three …, ten, twenty, thirty …, one hundred, two hundred …, one thousand, how many?, count, count up / to / on / back, count in fours, eights, fifties, hundreds, more, less, many, few, tally, odd, even, every other, skip count, how many times?, multiple of, sequence, continue, predict, pattern, pair, rule, relationship

Monday

Give each pair of pupils 48 small interlocking cubes or building bricks or large squared paper cut into strips of four.

Ask pupils to make 12 sets of cubes, each with 4 cubes linked / grouped together.

Ask pupils to count together from 0–48, pointing to each cube as they count. Each time you reach a multiple of 4, pupils shout those numbers. Repeat, saying the multiples of 4 out loud and silently saying all the other numbers. Count forwards and backwards.

Tuesday

Use Monday's sets of cubes and give each pupil a 100 square.

Repeat Monday's activity but this time, when you say a multiple of 4, pupils circle that number on their 100 square. Practise counting both forwards and backwards.

Wednesday

Use Monday's sets of cubes and give each pupil or pair of pupils a number line from 0–50 or 0–100.

Repeat Monday's activity but this time, when you say a multiple of 4, pupils circle that number on their number line. Practise counting both forwards and backwards.

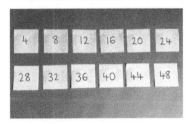

Thursday

Give each pupil or pair of pupils 12 sticky notes or small squares of paper.

Lay out the sticky notes. Nominate a starting point on the first square (e.g. bottom right-hand corner). Pupils count the corners on each sticky note and write the cumulative totals on each. Finish by counting forwards and backwards, using the numbers on the sticky notes as a resource.

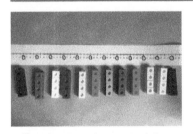

Friday

Use the sticky notes from Thursday.

In pairs, pupils lay out their sticky notes in order from 4–48. Show the pupils a playing card. Pupils count in fours that number of times and hold up the correct sticky note (e.g. you show an 8 and pupils hold up the sticky note for 8 × 4, i.e. 32).

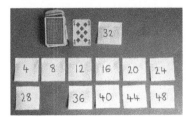

Week 2: Counting

Count from 0 in multiples of 8

Resources: objects, bowls, 100 squares, number lines, sticky notes, playing cards

Vocabulary: number, zero, one, two, three ..., ten, twenty, thirty ..., one hundred, two hundred ..., one thousand, how many?, count, count up / to / on / back, count in fours, eights, fifties, hundreds, more, less, many, few, tally, odd, even, every other, skip count, how many times?, multiple of, sequence, continue, predict, pattern, pair, rule, relationship

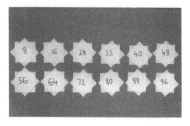

Monday

Give each pupil or pair of pupils a bowl of 100 objects (e.g. cubes, counters, buttons, pebbles, paper clips) or you could set a homework challenge to bring in a '100 jar'.

Ask pupils to make 12 sets, each with 8 objects grouped together. Pupils count together from 0–96, pointing to each object as they count. Each time you reach a multiple of 8, ask the pupils to shout those numbers. Count forwards and backwards.

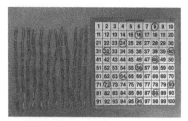

Tuesday

Use Monday's sets of objects. Give each pupil a 100 square.

Repeat Monday's activity but this time, when you say a multiple of 8, pupils circle that number on their 100 square. Practise counting both forwards and backwards.

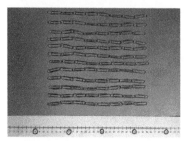

Wednesday

Use Monday's sets of objects. Give each pupil or pair of pupils a number line from 0–100.

Repeat Monday's activity but this time, when you say a multiple of 8, pupils circle that number on their number line. Practise counting both forwards and backwards.

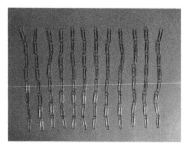

Thursday

Give each pupil or pair of pupils 24 sticky notes or small squares of paper.

Pupils create 12 stars, using two sticky notes for each eight-pointed star. Nominate a starting point on the first star (e.g. top point). Pupils count the points on each star and write the cumulative totals on each. Finish by counting forwards and backwards using the star resource.

Friday

Use Thursday's stars.

Pupils, individually or in pairs, lay out their stars in order from 8–96. Show pupils a playing card. Ask pupils to count in eights and to hold up the star which is the answer to the multiple shown on the playing card (e.g. you show a 6 and pupils hold up the star for 6 × 8, i.e. 48).

Week 3: Counting

Count from 0 in multiples of 4 and 8 *(relational understanding)*

Resources: number lines, 100 squares, playing cards

> **Vocabulary:** number, zero, one, two, three …, ten, twenty, thirty …, one hundred, two hundred …, one thousand, how many?, count, count up / to / on / back, count in fours, eights, fifties, hundreds, more, less, many, few, tally, odd, even, every other, skip count, how many times?, multiple of, sequence, continue, predict, pattern, pair, rule, relationship

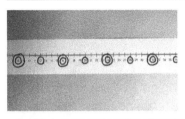

Monday

Give each pupil a number line from 0–100 (or ask them to draw one on their whiteboards or on long strips of paper). (Some pupils may benefit from a number line with numbers on rather than a blank one.)

Ask pupils to circle the first 12 multiples of 4 and the first 12 multiples of 8.

Tuesday

Give each pupil a 100 square.

Pupils count together in fours from 0–100, circling the multiples of 4 as they count. Repeat with multiples of 8. Two different colours could be used.

Ask pupils what they notice about the multiples of 4 and the multiples of 8. Can they form a 'rule' about the relationship between counting in fours and counting in eights (i.e. counting in eights is double counting in fours)?

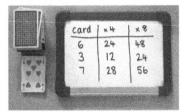

Wednesday

Give each pair of pupils a whiteboard and pen.

Pupils draw a grid, as shown. Show pupils a playing card and ask them to write that multiple of 4 and 8 on their whiteboards. Encourage them to use objects if they need to. Can they find an easy way to find the multiples of 8 (i.e. double the multiple of 4)?

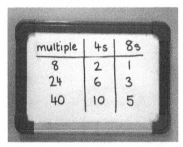

Thursday

Give each pair of pupils a whiteboard and pen.

Pupils draw a grid, as shown. Call out a multiple of 8 up to 48 (e.g. 40). Ask pupils to count together to that number, first in fours and then in eights, using their fingers to find the number of counting steps there are to the target number. Repeat. Can they find an easy way to find how many counting steps for eights (i.e. halve the number of steps for fours)?

Friday

Give each pupil or pair of pupils a whiteboard and pen.

Pupils draw a grid, as shown. Call out a multiple of 4 up to 48 (e.g. 44) and ask if we would say that number when counting in fours *and* eights. Count aloud to 48, first in fours and then in eights. Pupils write the number and tick or cross as appropriate.

Week 4: Counting

Count from 0 in multiples of 100

Resources: bundles of 100 objects, number lines, playing cards

Vocabulary: number, zero, one, two, three …, ten, twenty, thirty …, one hundred, two hundred …, one thousand, how many?, count, count up / to / on / back, count in fours, eights, fifties, hundreds, more, less, many, few, tally, odd, even, every other, skip count, how many times?, multiple of, sequence, continue, predict, pattern, pair, rule, relationship

Monday

Prepare a class set of resources, bundled in hundreds, to help pupils to count from 0–1000 in hundreds (e.g. Dienes, straws, cotton buds, pipe cleaners).

Count together from 0–1000 and back again, encouraging a counting rhythm and highlighting that when counting in hundreds, the tens and ones digits do not change. Repeat.

Display the 0–1000 resources for reference throughout the week.

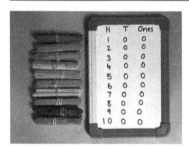

Tuesday

Use Monday's resources and give each pupil a whiteboard and pen.

Count together from 0–1000 in hundreds and back. Ask pupils to write the place-value headings H, T and Ones on their whiteboards. Again, count together forwards and backwards from 0–1000 in hundreds, with pupils recording the numbers on their whiteboards. Ask pupils what they notice about the pattern when counting in hundreds.

Wednesday

Give each pair of pupils an empty number line (or strips of paper or masking tape on the table tops).

Count together from 0–1000 in hundreds and back. Pupils mark 0 and 1000 on the number line (measuring the steps with a ruler may help). Ask pupils to write the multiples of 100 from 0–1000. Count backwards using the resource. Repeat to build fluency.

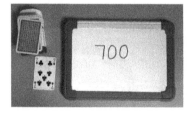

Thursday

Give each pupil a whiteboard and pen.

Show pupils a playing card (or digit card). Count together, in hundreds, that number of times. After a few goes, ask the pupils to do a 'counting check' before you start by writing the landing number (700 in this case) on a whiteboard. This will introduce the idea of strategies to check work.

Friday

Give each pair of pupils a pack of playing cards (or each pair could have a suit).

Write a multiple of 100 on the board and read the number out loud. Ask pupils to mentally count in hundreds to find the number of counting steps of 100 there are to reach the target number and then to show you a matching playing card.

Week 5: Counting

Count from 0 in multiples of 50

Resources: bundles of 50 objects, number lines, playing cards

> **Vocabulary:** number, zero, one, two, three ..., ten, twenty, thirty ..., one hundred, two hundred ..., one thousand, how many?, count, count up / to / on / back, count in fours, eights, fifties, hundreds, more, less, many, few, tally, odd, even, every other, skip count, how many times?, multiple of, sequence, continue, predict, pattern, pair, rule, relationship

Monday

Prepare a set of resources, bundled in fifties, to help pupils to count from 0–500 (e.g. Dienes, straws, cotton buds, pipe cleaners). Alternatively, one object could represent 10, as shown here.

Count together from 0–500 in fifties and back again, encouraging a counting rhythm and highlighting the patterns when counting. Repeat.

Tuesday

Create a second class set of a further 10 bundles of 50, enabling counting to 1000. Give each pupil a whiteboard and pen.

Count together from 0–1000 in fifties, both forwards and backwards. If time allows, record the numbers on the board and ask pupils what they notice about the pattern when counting in fifties.

Display the 0–1000 resources for reference throughout the week.

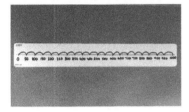

Wednesday

Give each pair of pupils an empty number line (or strips of paper or masking tape on the table tops).

Mark 0 and then count together from 0–1000 in fifties, asking pupils to mark the jumps as you go. Count backwards using the number line. Repeat to build fluency.

Thursday

Give each pupil a whiteboard and pen.

Show pupils a playing card. Count together in fifties that number of times and ask pupils to write the number on their whiteboard.

Friday

Give each pair of pupils a pack of playing cards (or each pair could have a suit).

Say a multiple of 50 (up to 500) out loud. Ask pupils to mentally count in fifties, using their fingers, to find the number of counting steps of 50 there are to reach the target number and then show you a matching playing card.

Week 6: Counting

Count from 0 in multiples of 50 and 100 *(relational understanding)*

Resources: number lines, playing cards

Vocabulary: number, zero, one, two, three ..., ten, twenty, thirty ..., one hundred, two hundred ..., one thousand, how many?, count, count up / to / on / back, count in fours, eights, fifties, hundreds, more, less, many, few, tally, odd, even, every other, skip count, how many times?, multiple of, sequence, continue, predict, pattern, pair, rule, relationship

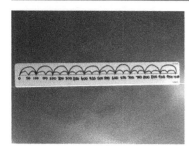

Monday

Give each pupil an empty number line from 0–1000 (or ask them to draw one on a whiteboard or on long strips of paper).

Ask pupils to draw the jumps from 0–1000 in hundreds on the number line. Repeat for the jumps of 50 within the hundreds. Use the number line to count together forwards and backwards in fifties and hundreds.

Ask pupils what they notice about the multiples of 50 and 100. Can they form a 'rule' about the relationship between counting in fifties and in hundreds (i.e. double the number of hundreds to find the number of fifties)?

Tuesday

Give each pair of pupils a whiteboard and pen.

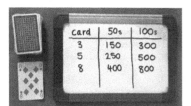

Pupils draw a grid, as shown, on their whiteboards. Call out a number between 1 and 9 (or show a playing card or digit card). Pupils count that number of times in fifties and then hundreds and record the numbers they reach.

Wednesday

Repeat Tuesday's activity, with pupils working individually.

Can they use the multiples of 50 to find multiples of 100?

Thursday

Give each pair of pupils a whiteboard and pen.

Pupils draw a grid, as shown, on their whiteboards. Call out a multiple of 100 up to 1000 (e.g. 700) and ask pupils to count together to that number, firstly in hundreds and then in fifties, using their fingers to find the number of counting steps there are to the target number. Repeat.

Can pupils think of an easy way to find how many counting steps there are for fifties (i.e. double the number of steps for hundreds to find the number of steps for fifties)?

Friday

Give each pupil or pair of pupils a whiteboard and pen.

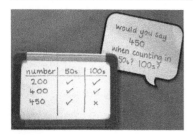

Pupils draw a grid, as shown, on their whiteboards. Call out a multiple of 50 up to 1000 (e.g. 450) and ask if we would say that number when counting in fifties *and* hundreds. Count aloud to 450, first in fifties and then in hundreds. Pupils write the number and tick or cross as appropriate.

Week 1: Place value

Recognise the place value of each digit in a three-digit number

Resources: place-value counters / Dienes

> **Vocabulary:** place value, place, ones, tens, hundreds, digit, one-, two-, three-digit number, 'teen' numbers, represents, exchange, the same as, equal to, greater, more, larger, less, fewer, smaller, greatest, most, largest, least, fewest, smallest, one more, ten more, one hundred more, one less, ten less, one hundred less, compare, order, first, second, third ..., last

---

Monday

Give each pupil some place-value counters and a whiteboard and pen.

Pupils draw a place-value grid, as shown, on their whiteboards. Call out a range of three-digit numbers and ask pupils to place counters to represent the number of hundreds, tens and ones. (Dienes is a good resource to show the proportionality of hundreds, tens and ones.)

Tuesday

Give each pupil some place-value counters and a whiteboard and pen.

Pupils draw a place-value grid, as shown, on their whiteboards. Call out a range of three-digit numbers that include zeros in the tens or ones columns and ask pupils to place counters to represent the number of hundreds, tens and ones.

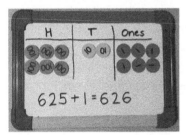

Wednesday

Give each pupil some place-value counters and a whiteboard and pen.

Pupils draw a place-value grid, as shown. Call out a three-digit number. Pupils add one more and place counters to represent the number of hundreds, tens and ones. Ask pupils to write the number sentence.

Repeat, this time finding one less.

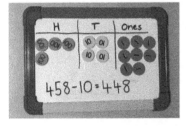

Thursday

Give each pupil some place-value counters and a whiteboard and pen.

Pupils draw a place-value grid, as shown. Call out a three-digit number. Pupils add ten more and place counters to represent the number of hundreds, tens and ones. Ask pupils to write the number sentence.

Repeat, this time finding ten less.

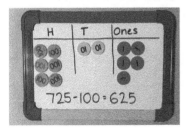

Friday

Give each pupil some place-value counters and a whiteboard and pen.

Pupils draw a place-value grid, as shown. Call out a three-digit number. Pupils add one hundred more and place counters to represent the number of hundreds, tens and ones. Ask pupils to write the number sentence.

Repeat, this time finding one hundred less.

Week 2: Place value

Compare and order numbers up to 1000

Resources: place-value counters / Dienes, number lines

> **Vocabulary:** place value, place, ones, tens, hundreds, digit, one-, two-, three-digit number, 'teen' numbers, represents, exchange, the same as, equal to, greater, more, larger, bigger, less, fewer, smaller, greatest, most, biggest, largest, least, fewest, smallest, one more, ten more, one hundred more, one less, ten less, one hundred less, compare, order, first, second, third …, last

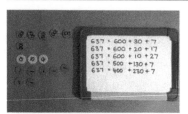

Monday

Give each pupil some place-value counters and a whiteboard and pen.

Say a three-digit number (e.g. 637). Ask pupils to make the number using place-value counters and then to record at least six different ways of partitioning that number.

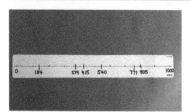

Tuesday

Give each pupil an empty number line (or a strip of paper or use masking tape on the desk tops).

Ask pupils to mark intervals for the hundreds from 0–1000. Call out a three-digit number and ask pupils to mark that number on the number line. Repeat.

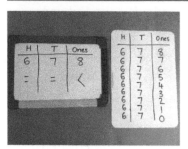

Wednesday

Give each pupil a whiteboard and pen.

Pupils draw a place-value grid, as shown, on their whiteboards. Call out a three-digit number and ask pupils to record the number. On the board, write the '=' symbol under the hundreds column, the '=' symbol under the tens column and the '<' symbol under the ones column. Pupils write all the possibilities for a new three-digit number that matches these conditions. Repeat with different three-digit numbers.

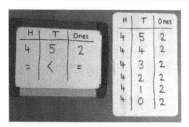

Thursday

Repeat Wednesday's activity, but write the '=' symbol under the hundreds column, the '<' symbol under the tens column and the '=' symbol under the ones column.

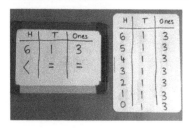

Friday

Repeat Wednesday's activity, but write the '<' symbol under the hundreds column, the '=' symbol under the tens column and the '=' symbol under the ones column.

Week 3: Place value

Read and write numbers up to 1000 in numerals

Resources: no additional resources required

Vocabulary: place value, place, ones, tens, hundreds, digit, one-, two-, three-digit number, 'teen' numbers, represents, exchange, the same as, equal to, greater, more, larger, bigger, less, fewer, smaller, greatest, most, biggest, largest, least, fewest, smallest, one more, ten more, one hundred more, one less, ten less, one hundred less, compare, order, first, second, third ..., last

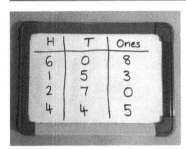

Monday

Give each pupil a whiteboard and pen.

Pupils draw a place-value grid, as shown. Say or show a three-digit number and ask pupils to record the number on their whiteboards. Ensure some numbers contain zeros (e.g. 270).

Say numbers in their expanded form but not in their place-value order (e.g. 4 tens, 5 ones and 4 hundreds), and then ask pupils to say the number.

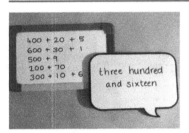

Tuesday

Give each pupil a whiteboard and pen.

Say a three-digit number (e.g. 316) and ask pupils to record the number on their whiteboards in expanded form (e.g. 300 + 10 + 6). Repeat, ensuring some numbers contain zeros.

Say numbers in expanded form but not in place-value order, and then ask pupils to say the number.

Wednesday

Give each pupil a whiteboard and pen.

Write a three-digit number on the board in numerals and ask pupils to write the number in words. Ensure some numbers contain zeros. Repeat.

Thursday

Give each pupil a whiteboard and pen.

Ask pupils to separate their whiteboards into three sections, as shown. Say a three-digit number. Pupils write the number in standard written form, in expanded form and in words. Ensure some numbers contain zeros. Repeat.

Friday

Give each pupil a whiteboard and pen.

Ask pupils to draw a grid, as shown. This is an opportunity for pupils to spot errors and check for accuracy. Write a three-digit number on the board in numerals (e.g. 816). Pupils copy the number on to their whiteboards. Then you say the number in words. If you say it correctly, the pupils tick the number they have written. If you do not say it correctly, they put a cross and write the number you said. Repeat.

Week 4: Place value

Read and write numbers up to 1000 in words

Resources: Dienes, place-value counters, 1–6 dice

Vocabulary: place value, place, ones, tens, hundreds, digit, one-, two-, three-digit number, 'teen' numbers, represents, exchange, the same as, equal to, greater, more, larger, bigger, less, fewer, smaller, greatest, most, biggest, largest, least, fewest, smallest, one more, ten more, one hundred more, one less, ten less, one hundred less, compare, order, first, second, third …, last

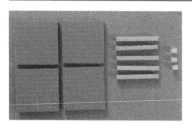

Monday

Give each pupil a whiteboard and pen.

Represent a three-digit number using Dienes (use a visualiser, hold up Dienes so all pupils can see, or use representations on an interactive screen). Pupils write the number in words and show you. Repeat.

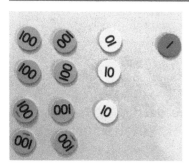

Tuesday

Give each pupil a whiteboard and pen.

Represent a three-digit number using place-value counters (use a visualiser, magnetic place-value counters or place-value counters attached to the board so all pupils can see, or representations on an interactive screen). Pupils write the number in words and show you. Repeat.

Wednesday

Give each pupil a whiteboard and pen.

Say a three-digit number. Pupils record the number on their whiteboards in word form. Ensure some numbers contain zeros. Repeat, saying numbers not in their place-value order (e.g. thirty, six hundred and five).

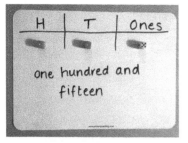

Thursday

Give each pair of pupils three dice and a whiteboard and pen.

Pupils draw a place-value grid, as shown, on their whiteboards. Pupils roll the dice and place them on the place-value grid. They then write the number in word form. Repeat.

Friday

Give each pupil a whiteboard and pen.

Write a three-digit number on the board in words. Ask pupils to write the number in numerals on their whiteboards. Repeat, ensuring some numbers contain zeros.

Week 5: Place value

Solve number problems using place-value ideas

Resources: 0–9 digit cards, counters, 1–6 dice

Vocabulary: place value, place, ones, tens, hundreds, digit, one-, two-, three-digit number, 'teen' numbers, represents, exchange, the same as, equal to, greater, more, larger, bigger, less, fewer, smaller, greatest, most, biggest, largest, least, fewest, smallest, one more, ten more, one hundred more, one less, ten less, one hundred less, compare, order, first, second, third …, last

Monday

Give each pair of pupils a set of digit cards.

Pupils take three cards and lay them out to make a three-digit number.

Partner 1 in each pair finds the largest number that can be made using the cards (932) while partner 2 finds the smallest number (239). Pupils check each other's work. Repeat.

Tuesday

Repeat Monday's activity, with pupils swapping roles.

Wednesday

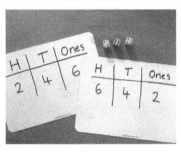

Give each pair of pupils 15 counters and a whiteboard and pen.

Tell pupils that you have used 15 counters to make a three-digit number on your place-value grid. Pupils draw a place-value grid on their whiteboards and try to work out the number you could have made. Pupils could record their answers using numbers, expanded form or drawings. Can they find all possibilities? Can they order the numbers from smallest to greatest?

Thursday

Give each pair of pupils three dice and a whiteboard and pen each.

Ask pupils to draw a place-value grid on their whiteboards. Partner 1 in each pair rolls the three dice and places them on the grid, saying the number out loud. Partner 1 records the largest number and partner 2 records the smallest number that can be made with the dice. Pupils check each other's work, then swap roles.

Friday

Give each pupil a whiteboard and pen.

On the board, write a three-digit number in expanded form with either the hundreds, tens or ones value missing, as shown. Say the number and ask pupils to write it in expanded form and to show you their whiteboards with the missing number written correctly.

Week 6: Place value

Solve practical problems using place-value ideas

Resources: A4 paper, counters, 1–6 dice

Vocabulary: place value, place, ones, tens, hundreds, digit, one-, two-, three-digit number, 'teen' numbers, represents, exchange, the same as, equal to, greater, more, larger, bigger, less, fewer, smaller, greatest, most, biggest, largest, least, fewest, smallest, one more, ten more, one hundred more, one less, ten less, one hundred less, compare, order, first, second, third …, last

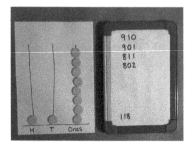

Monday

Give each pair of pupils a piece of A4 paper and 10 counters (or buttons, beads, etc.).

Ask pupils to draw an abacus model, as shown. Ask pupils to make the smallest three-digit number they can by placing all 10 counters on their abacus model.

Repeat for the greatest number and then for as many three-digit numbers as they can find. Pupils could record all the numbers and order them from smallest to largest.

Tuesday

Repeat Monday's activity with 20 counters.

Wednesday

Give each pair of pupils three dice and a whiteboard and pen each.

Pupils each draw a place-value grid, as shown, on their whiteboards. They each roll their three dice to create a three-digit number. The pupil who can create the greatest three-digit number from their dice wins. Repeat. Pupils keep a tally of their wins on their whiteboards.

Thursday

Repeat Wednesday's activity, but this time the pupil with the smallest three-digit number wins.

Gather the results from Wednesday and Thursday to find the greatest number and lowest number winners.

Friday

Give each pair of pupils three dice and a whiteboard and pen each.

Explain that the number shown by the dice will represent pence and pounds. Remind pupils that 100 pence are equal to one pound. Partner 1 in each pair rolls the three dice: the first dice represents hundreds, the second represents tens and the third represents ones. Partner 1 says and writes the number in pence. Partner 2 says and writes the number in pounds. Pupils check each other's work and then swap roles and repeat.

Week 1: Representing numbers

Identify, represent and estimate numbers using different representations *(place value)*

Resources: Dienes sets, place-value arrow cards

Vocabulary: place value, place, ones, tens, hundreds, digit, one-, two-, three-digit number, 'teen' numbers, represents, exchange, the same as, equal to, greater, more, larger, bigger, less, fewer, smaller, greatest, most, biggest, largest, least, fewest, smallest, one more, ten more, one hundred more, one less, ten less, one hundred less, compare, order, first, second, third …, last, estimate, nearly, roughly, close to, approximate, exactly, too many, too few, round up / down / to, nearest

Monday

Give each pair of pupils a set of Dienes. (If resources are limited, pupils could draw the Dienes on a place-value grid or use place-value counters.)

Show a three-digit number using place-value arrow cards. Pupils use Dienes to represent the number. Repeat.

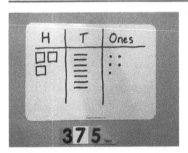

Tuesday

Give each pupil a whiteboard and pen.

Show a three-digit number in its partitioned form using place-value arrow cards. Pupils represent the number on a place-value grid. Repeat.

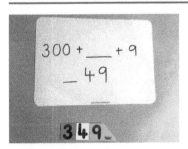

Wednesday

Give each pair of pupils a set of place-value arrow cards.

Draw a representation of a three-digit number on the board as shown. Pupils show that number using the place-value arrow cards. Ensure pupils take turns to make the numbers.

Thursday

Give each pair of pupils a set of place-value arrow cards.

Write a three-digit number on the board in both expanded and written numeral form with some digits missing in different places, as shown. Pupils use their place-value arrow cards to show the three-digit number. Repeat.

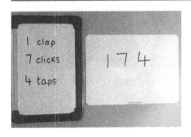

Friday

Give each pupil a whiteboard and pen.

Explain that claps represent hundreds, clicks of the fingers represent tens and a tap on the knees represents ones. Make a number using this body percussion. Pupils write the number, show you their whiteboards and say the number out loud. Repeat, with the representations not always in place-value order (e.g. 7 clicks, then 4 taps, then 1 clap for 174).

Week 2: Representing numbers

Identify, represent and estimate numbers using different representations *(place value)*

Resources: counters, 0–9 digit cards / playing cards

Vocabulary: place value, place, ones, tens, hundreds, digit, one-, two-, three-digit number, 'teen' numbers, represents, exchange, the same as, equal to, greater, more, larger, bigger, less, fewer, smaller, greatest, most, biggest, largest, least, fewest, smallest, one more, ten more, one hundred more, one less, ten less, one hundred less, compare, order, first, second, third …, last, estimate, nearly, roughly, close to, approximate, exactly, too many, too few, round up / down / to, nearest

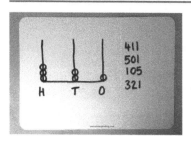

Monday

Give each pupil some counters and a whiteboard and pen.

Draw an abacus on the board as shown. Explain that you used nine beads, but that the beads on one of the columns have fallen off. Ask pupils to draw an abacus model and to use counters to represent all the beads, including the missing ones. Repeat.

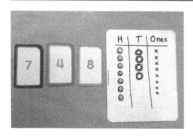

Tuesday

Repeat Monday's activity but, this time, explain that you used 12 beads and that some beads (which could be from any of the columns) have fallen off. Pupils explore all possibilities and record the numbers.

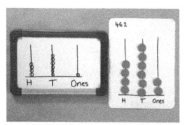

Wednesday

Give each pupil a whiteboard and pen.

Explain that, if you put three beads onto a tens / ones abacus, you could make the numbers 3, 30, 12 or 21. Then tell pupils that they have six counters which they can place anywhere on a hundreds, tens and ones abacus. What numbers can be made? Explore all possibilities.

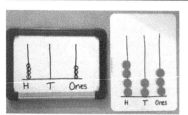

Thursday

Give each pair of pupils a whiteboard and pen.

Choose three digit cards or playing cards to represent hundreds, tens and ones and make a three-digit number. Say the number together. Pupils then create their own way of representing the number using pictures.

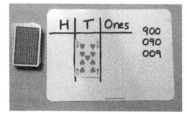

Friday

Give each pupil a whiteboard and pen.

Pupils draw a place-value grid, as shown, on their whiteboards. Display a digit card or playing card. Explain that pupils can only use that digit once to create numbers up to three digits. The remainder of the 'places' should be filled with zeros. Explore all possibilities. Explain that zeros to the left of the highest value do not represent any value (i.e. 009 equals 9). Repeat.

Week 3: Representing numbers

Identify, represent and estimate numbers using different representations *(money)*

Resources: place-value counters, bowls

Vocabulary: place value, place, ones, tens, hundreds, digit, one-, two-, three-digit number, 'teen' numbers, represents, exchange, the same as, equal to, greater, more, larger, bigger, less, fewer, smaller, greatest, most, biggest, largest, least, fewest, smallest, one more, ten more, one hundred more, one less, ten less, one hundred less, compare, order, first, second, third …, last, estimate, nearly, roughly, close to, approximate, exactly, too many, too few, round up / down / to, nearest

Monday

Give each pupil a whiteboard and pen.

Explain that all the price tags in a shop have been written in pence and that pupils need to write new labels in pounds. Remind pupils that one hundred pence are equal to one pound. Write an amount in pence on the board. Pupils write the amount using £ and p. Repeat.

Tuesday

Give each pupil a whiteboard and pen.

Explain that today all of the price tags have been written in £ and p, but the shopkeeper would like them all converted to pence only. Write an amount in £ and p on the board. Pupils write the amount using just pence. Repeat.

Wednesday

Give each pair of pupils a bowl of place-value counters and a whiteboard and pen.

Write an amount in pence on the board. Ask pupils to write this on their whiteboards. Challenge them to use counters to add 10p to the amount, exchanging as needed when crossing the hundreds boundary. Repeat.

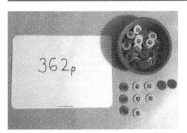

Thursday

Give each pair of pupils a bowl of place-value counters and a whiteboard and pen.

Explain that the amount you write on the board has already had 10p added to it. Ask pupils to use place-value counters to represent the original amount.

Friday

Repeat Thursday's activity but, this time, explain that the amount you write on the board has already had 100p (£1) added to it. Ask pupils to use place-value counters to represent the original amount.

Week 4: Representing numbers

Identify, represent and estimate numbers using different representations *(number lines)*

Resources: number lines

> **Vocabulary:** place value, place, ones, tens, hundreds, digit, one-, two-, three-digit number, 'teen' numbers, represents, exchange, the same as, equal to, greater, more, larger, bigger, less, fewer, smaller, greatest, most, biggest, largest, least, fewest, smallest, one more, ten more, one hundred more, one less, ten less, one hundred less, compare, order, first, second, third …, last, estimate, nearly, roughly, close to, approximate, exactly, too many, too few, round up / down / to, nearest

Monday

Give each pupil an empty, or partially scaled, number line.

Ask pupils to write 0 at one end and 1000 at the other end. Call out a three-digit number that is a multiple of 100 (to enable pupils to understand the intervals). Ask pupils to mark the number. Repeat.

Tuesday

Repeat Monday's activity but, this time, call out a three-digit number that is a multiple of 50 (to enable pupils to understand the intervals).

Wednesday

Repeat Monday's activity but, this time, call out a three-digit number that is a multiple of 10 (to enable pupils to understand the intervals).

Thursday

Give each pupil an empty, or partially scaled, number line and a whiteboard and pen.

Write a three-digit number on the board. Ask pupils to write the previous multiple of ten (750 in this case) and the next multiple of ten (760 in this case) on their whiteboard. Pupils first mark these numbers on their number line and then mark the number you have written. Repeat.

Friday

Give each pupil an empty, or partially scaled, number line.

Write a three-digit number on the board. Ask pupils to write the hundred before and the hundred after on their whiteboard and then on their number line. Then ask pupils to mark the number you have written. Repeat.

Week 5: Representing numbers

Identify, represent and estimate numbers using different representations *(scaling)*

Resources: cubes

Vocabulary: place value, place, ones, tens, hundreds, digit, one-, two-, three-digit number, 'teen' numbers, represents, exchange, the same as, equal to, greater, more, larger, bigger, less, fewer, smaller, greatest, most, biggest, largest, least, fewest, smallest, one more, ten more, one hundred more, one less, ten less, one hundred less, compare, order, first, second, third …, last, estimate, nearly, roughly, close to, approximate, exactly, too many, too few, round up / down / to, nearest

Monday

Give each pair of pupils 24 cubes.

Ask pupils to show you an amount of cubes which is twice as long as one you show them. Repeat using different cube totals up to 12. This may be pupils' first introduction to scaling, so modelling the concept of 'twice as many' is important: count 6 cubes, then 12 cubes.

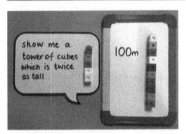

Tuesday

Give each pair of pupils 30 cubes and a whiteboard and pen.

Explain that today each cube represents 10 metres. Show a tower of cubes between 5 and 15 cubes in height. Ask pupils to show you what twice the height would be. Partner 1 in each pair makes the cube tower, while partner 2 writes the height on their whiteboard. Repeat, with pupils swapping roles.

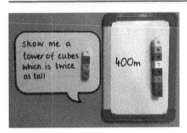

Wednesday

Give each pair of pupils 20 cubes and a whiteboard and pen.

Explain that today each cube represents 50 metres. Show a tower of cubes between 5 and 10 cubes in height. Ask pupils to show you what twice the height would be. Partner 1 in each pair makes the cube tower, while partner 2 writes the height on their whiteboard. Repeat, with pupils swapping roles.

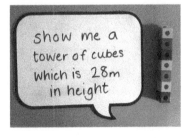

Thursday

Give each pupil 20 cubes and a whiteboard and pen.

Explain that today each cube represents 4 metres. Tell pupils that you are going to give a tower height and they have to make a tower of that height, with each cube representing 4 metres, and write the height on their whiteboard. Once towers have been built, count forwards and backwards in multiples of 4 metres, pointing at the cubes for each counting step. Repeat.

Friday

Give each pupil 12 cubes and a whiteboard and pen.

Explain that today each cube represents 8 metres. Tell pupils that you are going to give a tower height and they have to make a tower of that height, with each cube representing 8 metres, and write the height on their whiteboard. Once towers have been built, count forwards and backwards in multiples of 8 metres, pointing at the cubes for each counting step. Repeat.

Week 6: Representing numbers

Identify, represent and estimate numbers using different representations (partitioning)

Resources: place-value arrow cards, cubes

Vocabulary: place value, place, ones, tens, hundreds, digit, one-, two-, three-digit number, 'teen' numbers, represents, exchange, the same as, equal to, greater, more, larger, bigger, less, fewer, smaller, greatest, most, biggest, largest, least, fewest, smallest, one more, ten more, one hundred more, one less, ten less, one hundred less, compare, order, first, second, third …, last, estimate, nearly, roughly, close to, approximate, exactly, too many, too few, round up / down / to, nearest

Monday

Give each pair of pupils a set of place-value arrow cards and a whiteboard and pen.

Explain that today they will be partitioning into hundreds, tens and ones. Say a three-digit number. Partner 1 in each pair makes the number using place-value arrow cards while partner 2 writes the number using expanded (partitioned) form. Repeat with pupils swapping roles.

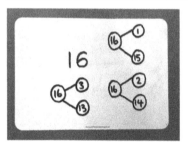

Tuesday

In advance of Wednesday's activity, pupils explore a part / whole partitioning model. It is important that pupils have been exposed to HTO partitioning, as in Monday's activity, as well as part / whole models.

Write a teen number on the board. Explore partitioning the number in all ways and recording these systematically. Draw a model to support understanding.

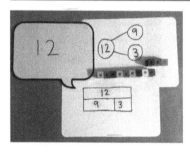

Wednesday

Give each pupil 20 cubes and a whiteboard and pen.

Say a two-digit number (up to 20) and model using the two partitioning methods shown. Also model using the cubes to help secure understanding of partitioning.

Pupils practise creating both models on their whiteboards. Encourage using the cubes to help secure understanding of partitioning.

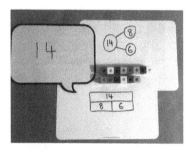

Thursday

Give each pair of pupils 20 cubes and a whiteboard and pen each.

Say a teen number and model using the two partitioning methods shown.

Practise with partner 1 in each pair creating the satellite (circles) model and partner 2 creating the bar model on their whiteboards. Encourage using the cubes to help secure understanding of partitioning.

Friday

Repeat Thursday's activity, with pupils swapping roles.

Week 1: Addition and subtraction

Add mentally a three-digit number and ones

Resources: dice, objects

Vocabulary: digit, add, addition, more, plus, make, sum, total, altogether, one more, two more ... ten more ... one hundred more, how many more to make ...?, missing number, how many more is ...?, how much more is ...?, subtract, subtraction, take away, minus, leave, how many are left / left over?, one less, two less ... ten less ... one hundred less, how many fewer?, how much less?, difference between, +, −, =, equals, sign, is the same as, boundary, exchange

Monday

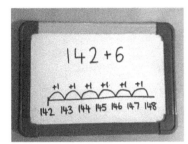

Give each pair of pupils a whiteboard and pen.

Write 142 + 6 on the board. Ask pupils to discuss how they will solve this. Record their ideas on the board. Display throughout the week.

Model drawing a number line and counting on 6 ones from 142. Repeat with other three-digit numbers, adding ones (not crossing the tens boundary).

Tuesday

Give each pair of pupils a dice and a whiteboard and pen.

Start by modelling the number line (counting on) method as in Monday's activity. Pairs follow the model on their whiteboards.

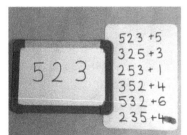

Write three digits on the board. Ask pupils to use the digits to generate six different three-digit numbers. Partner 1 in each pair rolls the dice and adds the number shown on the dice to the numbers they have created. Partner 2 checks the answers by counting along the number line.

Wednesday

Repeat Tuesday's activity, with pupils swapping roles.

Thursday

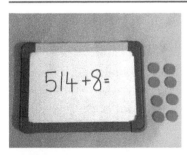

Give each pair of pupils 10 objects (e.g. counters or cubes).

Write '514 + 8 =' on the board and say the calculation out loud together. Model counting out 6 objects to get to the nearest ten, and then 2 more. So, 514 + 8 = 514 + 6 + 2. Allow pairs time to practise with other calculations that cross the tens (e.g. 223 + 8, 657 + 6, 325 + 7, 423 + 9).

Friday

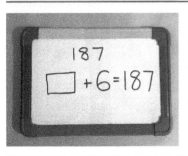

Write 187 on the board and explain that 6 has already been added to give this answer. Write '☐ + 6 = 187' on the board. Ask what number goes in the box, challenging pupils to explain how they calculated this (e.g. bar model, count back / subtraction strategies, a 'balance' model). Repeat with other calculations taking the form ☐ + x = y.

Week 2: Addition and subtraction

Subtract mentally a three-digit number and ones

Resources: dice, objects

> **Vocabulary:** digit, add, addition, more, plus, make, sum, total, altogether, one more, two more … ten more … one hundred more, how many more to make…?, missing number, how many more is …?, how much more is …?, subtract, subtraction, take away, minus, leave, how many are left / left over?, one less, two less … ten less … one hundred less, how many fewer?, how much less?, difference between, +, −, =, equals, sign, is the same as, boundary, exchange

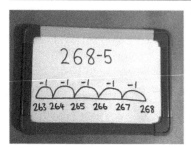

Monday

Give each pair of pupils a whiteboard and pen.

Write 268 – 5 on the board. Ask pupils to discuss how they will solve this. Record their ideas on the board. Display throughout the week.

Model drawing a number line and counting back 5 ones from 268. Repeat with other three-digit numbers, subtracting ones (not crossing the tens boundary).

Tuesday

Give each pair of pupils a dice and a whiteboard and pen.

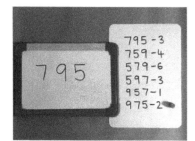

Start by modelling the number line (counting back) method as in Monday's activity. Pairs follow the model on their whiteboards.

Write three digits on the board. Ask pupils to use the digits to generate six different three-digit numbers. Partner 1 in each pair rolls the dice and subtracts the number shown on the dice from the numbers they have created. Partner 2 checks the answers by counting along the number line.

Wednesday

Repeat Tuesday's activity, with pupils swapping roles.

Thursday

Give each pair of pupils 10 objects (e.g. counters or cubes).

Write 364 – 7 on the board and say the calculation out loud together. Model removing 4 counters to get to the nearest ten, and then 3 more. So, 364 – 7 = (364 – 4) – 3. Model the same calculation using fingers to count back. Allow pairs time to practise with other calculations that cross the tens (e.g. 523 – 6, 413 – 9, 478 – 9, 615 – 7).

Friday

Write 251 on the board and explain that 5 has already been subtracted to give this answer. Write '☐ – 5 = 251' on the board. Ask what number goes in the box, challenging pupils to explain how they calculated this (e.g. bar model, count back / subtraction strategies, a 'balance' model). Repeat with other calculations taking the form $\boxed{} - x = y$.

Week 3: Addition and subtraction

Add mentally a three-digit number and tens

Resources: place-value counters / normal counters / Dienes, counting stick, cubes

Vocabulary: digit, add, addition, more, plus, make, sum, total, altogether, one more, two more … ten more … one hundred more, how many more to make …?, missing number, how many more is …?, how much more is …?, subtract, subtraction, take away, minus, leave, how many are left / left over?, one less, two less … ten less … one hundred less, how many fewer?, how much less?, difference between, +, −, =, equals, sign, is the same as, boundary, exchange

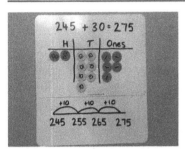

Monday

Give each pair of pupils some place-value counters (or normal counters or Dienes) and a whiteboard and pen.

Write 245 + 30 on the board. Ask pupils to discuss how they will solve this. Record their ideas on the board. Display throughout the week.

Model drawing a place-value grid and adding 3 tens. Then draw a number line to represent the same calculation. Repeat with other calculations (e.g. 326 + 50, 615 + 40), not crossing the hundreds boundary at first.

Tuesday

Give each pupil 10 cubes. Ask pupils to make their own 10-cube counting stick.

Start by using the counting stick to count on in tens from 417.

Write 172 on the board and say the number together. Count forwards in tens from 172, then backwards to the start number. Repeat with other start numbers.

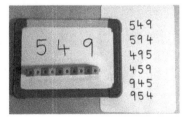

Wednesday

Give each pair of pupils a cube counting stick and a whiteboard and pen.

Write three digits on the board. Pupils generate all possible three-digit numbers and write them on their whiteboards. They use the counting stick to count forwards and backwards in tens, to and from each number they generate.

Thursday

Give each pair of pupils a cube counting stick and a whiteboard and pen.

Tell pupils that each cube now represents 10. Write 648 on the board. Break off 4 cubes and ask how many tens are represented (4). Ask the value of the cubes (40). Add 40 to 648. Count together, pointing at the cubes. Write '648 + 40 = 688' on the board. Repeat using the other 6 cubes.

Partner 1 in each pair writes a three-digit number. Partner 2 shows a quantity from their counting stick. Partner 1 writes both calculations and partner 2 checks the answers. Ask pupils which digits change. Do the ones change?

Friday

Repeat Thursday's activity, with pupils swapping roles.

Week 4: Addition and subtraction

Subtract mentally a three-digit number and tens

Resources: place-value counters / normal counters / Dienes, counting stick, cubes

> **Vocabulary:** digit, add, addition, more, plus, make, sum, total, altogether, one more, two more … ten more … one hundred more, how many more to make …?, missing number, how many more is …?, how much more is …?, subtract, subtraction, take away, minus, leave, how many are left / left over?, one less, two less … ten less … one hundred less, how many fewer?, how much less?, difference between, +, −, =, equals, sign, is the same as, boundary, exchange

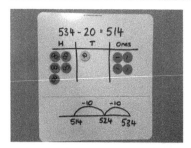

Monday

Give each pair of pupils some place-value counters and a whiteboard and pen.

Write 534 − 20 on the board. Ask pupils to discuss how they will solve this. Record their ideas on the board. Display throughout the week. Model drawing a place-value grid and subtracting 2 tens. Then draw a number line to represent the same calculation. Repeat with other calculations (e.g. 479 − 60, 971 − 50); start by not crossing the hundreds boundary and then move on to exchange.

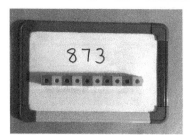

Tuesday

Give each pupil 10 cubes.

Start by using a counting stick to count back in tens from 564 to 464.

Ask pupils to make their own 10-cube counting stick. Write 873 on the board and say the number together. Count backwards in tens from 873, then forwards to the start number. Repeat with other start numbers. Keep the cube counting sticks for use throughout the week.

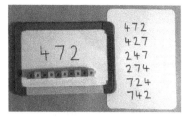

Wednesday

Give each pair of pupils a cube counting stick and a whiteboard and pen.

Write three digits on the board. Pupils generate all possible three-digit numbers and write them on their whiteboards. They use the counting stick to count backwards and then forwards in tens, from and to each number they generate.

Thursday

Give each pair of pupils a cube counting stick and a whiteboard and pen.

Tell pupils that each cube now represents 10. Write 396 on the board. Break off 8 cubes and ask how many tens are represented (8). Ask the value of the cubes (80). Subtract 80 from 396. Count together, pointing at the cubes. Write '396 − 80 = 316' on the board. Repeat, using the other 2 cubes.

Partner 1 in each pair writes a three-digit number. Partner 2 shows a quantity from their counting stick. Partner 1 writes both calculations and partner 2 checks the answers. Highlight using inverses to check answers. Ask pupils which digits change. Do the ones change?

Friday

Repeat Thursday's activity, with pupils swapping roles.

Week 5: Addition and subtraction

Add mentally a three-digit number and hundreds

Resources: place-value counters / normal counters / Dienes, counting stick, cubes

> **Vocabulary:** digit, add, addition, more, plus, make, sum, total, altogether, one more, two more … ten more … one hundred more, how many more to make …?, missing number, how many more is …?, how much more is …?, subtract, subtraction, take away, minus, leave, how many are left / left over?, one less, two less … ten less … one hundred less, how many fewer?, how much less?, difference between, +, −, =, equals, sign, is the same as, boundary, exchange

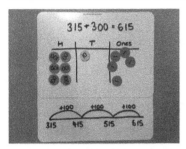

Monday

Give each pair of pupils some place-value counters (or normal counters or Dienes) and a whiteboard and pen.

Write 315 + 300 on the board. Ask pupils to discuss how they will solve this. Record their ideas on the board. Display throughout the week.

Model drawing a place-value grid on the board and adding 3 hundreds. Then draw a number line to represent the same calculation. Repeat with other calculations.

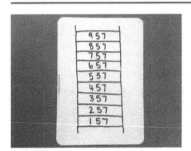

Tuesday

Give each pair of pupils a whiteboard and pen.

Ask pupils to draw a 'ladder' with nine spaces. Write a three-digit number with a 1 in the hundreds column in the bottom space. Count up in hundreds, recording the numbers in each space. When you have finished, count together forwards and backwards in hundreds from that number. Repeat for other start numbers under 199.

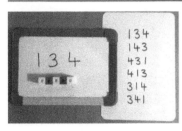

Wednesday

Give each pair of pupils a 5-cube counting stick, with each cube representing 100, and a whiteboard and pen.

Write 134 on the board. Pupils generate all possible three-digit numbers and write them on their whiteboards. They use the counting stick to count forwards and backwards, to and from each number they generate.

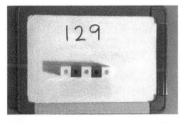

Thursday

Give each pair of pupils a 5-cube counting stick.

Write 129 on the board and say the number together. Count forwards and backwards in hundreds from 129. Repeat with other start numbers under 499.

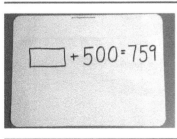

Friday

Write 759 on the board and explain that 500 has already been added to give this answer. Write '☐ + 500 = 759' on the board. Ask what number goes in the box. Repeat with other calculations taking the form $\square + x = y$. Link to inverses.

Week 6: Addition and subtraction

Subtract mentally a three-digit number and hundreds

Resources: place-value counters / normal counters / Dienes, counting stick, cubes

> **Vocabulary:** digit, add, addition, more, plus, make, sum, total, altogether, one more, two more … ten more … one hundred more, how many more to make …?, missing number, how many more is …?, how much more is …?, subtract, subtraction, take away, minus, leave, how many are left / left over?, one less, two less … ten less … one hundred less, how many fewer?, how much less?, difference between, +, −, =, equals, sign, is the same as, boundary, exchange

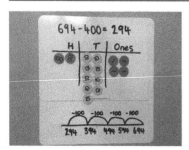

Monday

Give each pair of pupils some place-value counters (or normal counters or Dienes) and a whiteboard and pen.

Write 694 – 400 on the board. Ask pupils to discuss how they will solve this. Record their ideas on the board. Display throughout the week.

Model drawing a place-value grid on the board and subtracting 4 hundreds. Then draw a number line to represent the same calculation. Repeat with other calculations (e.g. 657 – 300, 759 – 600).

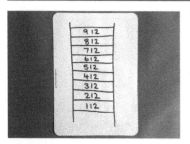

Tuesday

Give each pair of pupils a whiteboard and pen.

Ask pupils to draw a 'ladder' with nine spaces. Write a three-digit number with a 9 in the hundreds column in the top space. Count down in hundreds, recording the numbers in each space. When you have finished, count forwards and backwards in hundreds from that number. Repeat for other start numbers over 900.

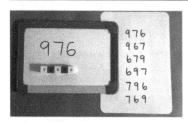

Wednesday

Give each pair of pupils a 5-cube counting stick, with each cube representing 100, and a whiteboard and pen.

Write 976 on the board. Pupils generate all possible three-digit numbers and write them on their whiteboards. They use the counting stick to count forwards and backwards in hundreds, from and to each number they generate.

Thursday

Give each pair of pupils a 5-cube counting stick, with each cube representing 100.

Write 837 on the board and say the number together. Count backwards and forwards in hundreds from 837. Repeat with other start numbers over 500 and different cube representations for counting.

Friday

Write 608 on the board and explain that 200 has already been subtracted to give this answer. Write '☐ – 200 = 608' on the board. Ask what number goes in the box. Repeat with other calculations taking the form ☐ – x = y. Link to inverses.

Week 1: Multiplication and division

Recall and use multiplication and division facts for the 3 multiplication table

Resources: counting stick / 100 square, cubes

Vocabulary: lots, groups of, times, multiply, multiplication, multiplied by, multiple of, product, once, twice, three times … ten times as big / long / wide …, repeated addition, array, row, column, double, halve, share, share equally, one each, two each, three each …, group in pairs, threes … twelves, equal groups of, divide, division, divided by / into, left, left over, remainder, ×, ÷

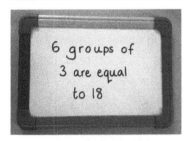

Monday

Start by counting in threes from 0–36 and back again, using either a counting stick or a 100 square to support.

Write '6 groups of 3 are equal to 18' on the board. Ask pupils to prove it in as many ways as they can (e.g. using pictures, number sentences, arrays, groups, objects, number lines for repeated addition). Take photographs to display throughout the week.

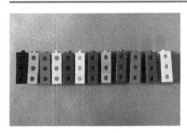

Tuesday

Give each pair of pupils 36 cubes.

Start by counting in threes from 0–36 and back again, using either a counting stick or a 100 square to support.

Pairs make 12 towers of 3 cubes. Count together in threes, forwards and backwards, putting down a tower of 3 cubes each time. Practise again, starting from different multiples of 3. Keep the towers to use throughout the week.

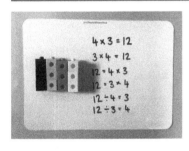

Wednesday

Give each pair of pupils a set of 12 towers of 3 cubes and a whiteboard and pen.

Start by counting in threes from 0–36 and back again, using either a counting stick or a 100 square to support.

Ask pairs to create an array from 4 towers of 3 cubes and to write as many facts as they can. Now ask how many columns of 3 there are in 12. Model writing '12 ÷ 3 = 4'. Next ask pupils to rotate the array to see how many columns of 4 there are in 12. Model writing facts.

Thursday

Give each pair of pupils a set of 12 towers of 3 cubes and a whiteboard and pen.

Call out a 3 multiplication table fact (e.g. 9 × 3). Partner 1 in each pair makes an array that demonstrates this. Partner 2 records all the multiplication and division facts relating to the array.

Friday

Repeat Thursday's activity, with pupils swapping roles.

Week 2: Multiplication and division

Recall and use multiplication and division facts for the 4 multiplication table

Resources: counting stick / 100 square, cubes

Vocabulary: lots, groups of, times, multiply, multiplication, multiplied by, multiple of, product, once, twice, three times … ten times as big / long / wide …, repeated addition, array, row, column, double, halve, share, share equally, one each, two each, three each …, group in pairs, threes … twelves, equal groups of, divide, division, divided by / into, left, left over, remainder, ×, ÷

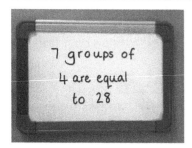

Monday

Start by counting in fours from 0–48 and back again, using either a counting stick or a 100 square to support. (Animals with four legs are a good resource for counting in fours as they give a practical context that the pupils know.)

Write '7 groups of 4 are equal to 28' on the board (could show 7 animals). Ask pupils to prove it in as many ways as they can (e.g. using pictures, number sentences, arrays, groups, objects, number lines for repeated addition). Take photographs to display throughout the week.

Tuesday

Give each pair of pupils 48 cubes.

Start by counting in fours from 0–48 and back again, using either a counting stick or a 100 square to support.

Pairs make 12 towers of 4 cubes. Count together in fours, forwards and backwards, putting down a tower of 4 cubes each time. Practise again, starting from different multiples of 4. Keep the towers to use throughout the week.

Wednesday

Give each pair of pupils a set of 12 towers of 4 cubes and a whiteboard and pen.

Start by counting in fours from 0–48 and back again, using either a counting stick or a 100 square to support.

Ask pairs to create an array from 6 towers of 4 cubes and to write as many facts as they can. Now ask how many columns of 4 there are in 24. Model writing '24 ÷ 4 = 6' on the board. Next ask pupils to rotate the array to see how many columns of 6 there are in 24. Model writing facts.

Thursday

Give each pair of pupils a set of 12 towers of 4 cubes and a whiteboard and pen.

Call out a 4 multiplication table fact (e.g. 8 × 4). Partner 1 in each pair makes an array that demonstrates this. Partner 2 records all the multiplication and division facts relating to the array.

Friday

Repeat Thursday's activity, with pupils swapping roles.

Week 3: Multiplication and division

Recall and use multiplication and division facts for the 8 multiplication table

Resources: counting stick / 100 square, objects, cubes

> **Vocabulary:** lots, groups of, times, multiply, multiplication, multiplied by, multiple of, product, once, twice, three times … ten times as big / long / wide …, repeated addition, array, row, column, double, halve, share, share equally, one each, two each, three each …, group in pairs, threes … twelves, equal groups of, divide, division, divided by / into, left, left over, remainder, ×, ÷

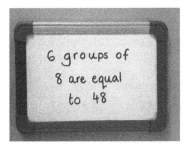

Monday

Start by counting in eights from 0–96 and back again, using either a counting stick or a 100 square to support. (Animals with four legs banded in twos are a good resource for showing how counting in eights relates to counting in fours.)

Write '6 groups of 8 are equal to 48' on the board. Ask pupils to prove it in as many ways as they can (e.g. using pictures, number sentences, arrays, groups, objects, number lines for repeated addition). Take photographs to display throughout the week.

This week, make available a good choice of different objects.

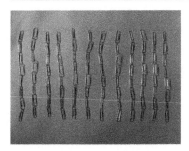

Tuesday

Give each group / pair of pupils 96 cubes to make 12 towers of 8 cubes or ask pupils to make their own set of 12 groups of 8 objects to use throughout the week.

Start by counting in eights from 0–96 and back again, using either a counting stick or a 100 square to support.

Count together in eights, forwards and backwards, touching one set of objects each time you count on. Practise again, starting from different multiples of 8.

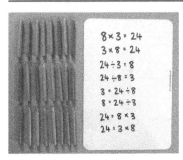

Wednesday

Give each pair of pupils their set of cubes or objects and a whiteboard and pen.

Start by counting in eights from 0–96 and back again, using either a counting stick or a 100 square to support.

Ask pairs to create an array from 3 of their groups of objects and to write as many facts as they can. Ask how many columns of 3 there are in 24. Model writing '24 ÷ 3 = 8'. Next ask pupils to rotate the array to see how many columns of 8 there are in 24. Model writing facts.

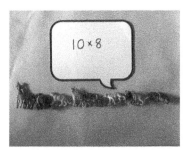

Thursday

Give each pair of pupils their set of cubes or objects and a whiteboard and pen.

Call out an 8 multiplication table fact (e.g. 10 × 8). Partner 1 in each pair creates a model, representation or array that demonstrates this. Partner 2 records all the multiplication and division facts relating to the array.

Friday

Repeat Thursday's activity, with pupils swapping roles.

Solve problems, including missing number problems, involving multiplication and division

Resources: bowls, objects, sticky notes / paper, hoops

Vocabulary: lots, groups of, times, multiply, multiplication, multiplied by, multiple of, product, once, twice, three times … ten times as big / long / wide …, repeated addition, array, row, column, double, halve, share, share equally, one each, two each, three each …, group in pairs, threes … twelves, equal groups of, divide, division, divided by / into, left, left over, remainder, ×, ÷

Monday

Place different quantities of objects in bowls in multiples of 3. Tell pupils that today's focus is multiplication and division facts within the 3 multiplication table.

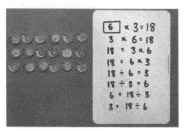

Working in pairs, pupils count the total number of objects in each bowl, then partner 1 in each pair places them in an array. Partner 2 writes ☐ × 3 = ☐. They then write a 3 multiplication fact and as many other facts as possible.

Tuesday

Repeat Monday's activity with pupils swapping roles.

Wednesday

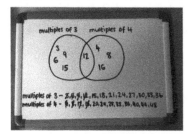

Give each pair of pupils a whiteboard and pen and ask them to draw a large Venn diagram, labelled as shown.

Partner 1 in each pair writes all the multiples of 3 up to 36 on the whiteboard. Partner 2 writes all the multiples of 4 up to 48 (as shown).

Ask pupils to take turns to write one of their numbers in the correct place on the Venn diagram.

Thursday

Give each pair of pupils 24 sticky notes or squares of paper and two hoops, or ask pupils to draw a Venn diagram on their whiteboard.

Partner 1 in each pair writes all the multiples of 4 up to 48. Partner 2 writes all the multiples of 8 up to 96.

Ask pupils to sort their numbers. Ask if there would be any multiples of 8 that are not multiples of 4.

Friday

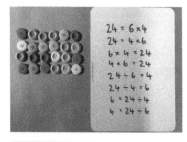

Give each pair of pupils a set of 24 objects and a whiteboard and pen.

Ask them to count out the objects. Tell pupils that they should arrange the objects to create as many arrays as possible using all 24 objects and then record as many multiplication and division facts as possible. (A 6 × 4 array is shown here, but pupils should repeat for 8 × 3 and 2 × 12).

Week 5: Multiplication and division

Solve problems involving multiplication, including positive integer scaling problems

Resources: place-value counters, bowls, counters, cubes

Vocabulary: lots, groups of, times, multiply, multiplication, multiplied by, multiple of, product, once, twice, three times … ten times as big / long / wide …, repeated addition, array, row, column, double, halve, share, share equally, one each, two each, three each …, group in pairs, threes … twelves, equal groups of, divide, division, divided by / into, left, left over, remainder, ×, ÷

Monday

Give each pair of pupils some place-value counters.

Explain that you have been given a recipe that makes a cake for 8 people, but you need to make enough for 16 people. Ask how they can solve this problem. Ask pupils to use place-value counters to show the quantities needed for a cake for 16 people.

Tuesday

Prepare a bowl with 6 blue, 3 red and 5 green counters. Give each pair of pupils a quantity of each colour counter.

Show the class the bowl of counters, counting out each colour in ones. Explain that pupils need to work in pairs to increase the quantities of counters so there are 3 times as many of each colour. (If there are not enough counters, pupils could draw pictures to represent their answer.)

Wednesday

Give each pupil 12 cubes and a whiteboard and pen.

Explain that today each cube represents 4 metres. Tell pupils that they will be given a height, in metres, and they must show you the correct number of cubes to represent a tower of that height. Write and say 'Show me a tower of 28 metres'. Once towers have been built, count forwards and backwards in multiples of 4m, pointing at the cubes for each counting step. Repeat for other multiples of 4.

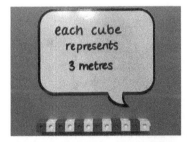

Thursday

Give each pupil 12 cubes and a whiteboard and pen.

Explain that today each cube represents 3 metres. Tell pupils that they must create towers of a specific height, with each cube representing 3m (e.g. a tower of 36m). Once towers have been built, count forwards and backwards in metres, pointing at the cubes for each counting step. Repeat for other multiples of 3.

Friday

Repeat Wednesday's activity but, this time, each cube represents 8 metres.

Week 6: Multiplication and division

Solve problems involving multiplication *(including commutativity)*

Resources: counters, dice

> **Vocabulary:** lots, groups of, times, multiply, multiplication, multiplied by, multiple of, product, once, twice, three times … ten times as big / long / wide …, repeated addition, array, row, column, double, halve, share, share equally, one each, two each, three each …, group in pairs, threes … twelves, equal groups of, divide, division, divided by / into, left, left over, remainder, ×, ÷

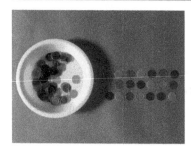

Monday

Give each pair of pupils some counters.

Ask pupils to create an array that shows 7 × 3 = 21. Remind pupils that the array can be rotated to show that 3 × 7 = 21. Write '3 × 7 = 7 × 3' on the board.

Repeat for an array showing 9 × 3.

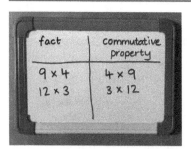

Tuesday

Give each pair of pupils some counters, two dice and a whiteboard and pen.

Ask pupils to roll the dice to generate a multiplication question. Partner 1 in each pair creates that array. Partner 2 then writes the number sentence for the multiplication fact they have created *and* the commutative property, as in Monday's activity.

Wednesday

Repeat Tuesday's activity, with pupils swapping roles.

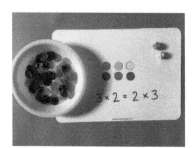

Thursday

Give each pupil a whiteboard and pen.

Draw or create an 8 × 3 array on the board. Ask pupils to write the number sentence and the commutative properties of the array. Ask if they can think of any other multiplication facts that are equal to 24.

Repeat with another array.

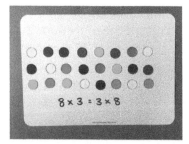

Friday

Give each pupil a whiteboard and pen.

Write '☐ × ☐ = 36' on the board. Ask pupils to draw a grid on their whiteboards as shown. They write one fact to solve the problem *and* one fact derived from commutativity.

Repeat for products of 12, 18, 24 and 30.

Week 1: Fractions

Count up and down in tenths

Resources: cubes, ten frames (e.g. printed, ice-cube trays, egg boxes)

Vocabulary: whole, part, equal parts, fraction, one whole, one half, two halves, one quarter, two quarters, three quarters, four quarters, one third, two thirds, three thirds, one tenth, two tenths … ten tenths, proportion, in every, for every, decimal, decimal fraction / point / place, numerator, denominator, equivalent, same, equal to

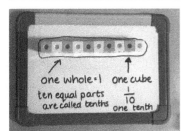

Monday

Give each pupil 10 cubes.

Ask pupils to make a cube snake. Explain that they have made one whole snake of 10 cubes. Tell them that the snake is divided into ten equal parts and that these are called tenths. Count forwards and backwards in tenths from 0–1. Repeat.

Tape a snake to the board and label the parts. Display throughout the week.

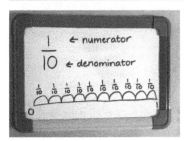

Tuesday

Give each pair of pupils a snake of 10 cubes and a whiteboard and pen.

Use the cube snakes to count forwards and backwards in tenths from 0–1, and then from any given tenth (i.e. not always starting at 0). Recap yesterday's vocabulary. Name and label the numerator and denominator. Pupils draw a number line and record the jumps of $\frac{1}{10}$ from 0–1.

Wednesday

Give each pupil a ten frame and 10 cubes.

Explain that they have one ten frame, which is divided into ten equal parts called tenths. Place one cube in each part and explain that 10 tenths are equal to one whole. Together count backwards in tenths from 1–0, removing one cube each time, and then forwards.

Thursday

Give each pupil a ten frame and 10 cubes.

Count forwards and backwards in tenths, from 0–1.

Call out a fraction as a tenth (e.g. six tenths). Ask pupils to represent it using cubes. Count up from 0 to that fraction and back to 0 again. Repeat for other tenths.

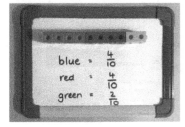

Friday

Prepare towers of 10 cubes, made up of two or three different colour cubes. Give each pupil a whiteboard and pen.

Show one tower and ask pupils to write and show you the fractions that are red, green and blue. Repeat for other towers.

Week 2: Fractions

Recognise, find and write fractions of a discrete set of objects

Resources: bowls, objects

Vocabulary: whole, part, equal parts, fraction, one whole, one half, two halves, one quarter, two quarters, three quarters, four quarters, one third, two thirds, three thirds, one tenth, two tenths … ten tenths, proportion, in every, for every, decimal, decimal fraction / point / place, numerator, denominator, equivalent, same, equal to

Monday

Prepare bowls of objects in multiples of 8 (e.g.16 cubes, 40 counters, 24 paper clips), enough for at least one bowl per pair, for use throughout the week. Give each pupil a whiteboard and pen.

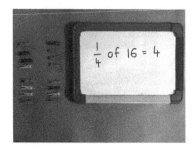

Write $\frac{1}{4}$ on the board. Explain that pupils will be finding $\frac{1}{4}$ of their set of objects. Remind pupils about the numerator and denominator and explain that, to find $\frac{1}{4}$, we share the objects into four equal parts. Model sharing into four equal groups. Pupils write the number sentence on their whiteboards.

Repeat with different sets of objects.

Tuesday

Repeat Monday's activity, this time with pupils finding $\frac{1}{8}$ of their set of objects.

Wednesday

Give each pair of pupils a set of objects in a multiple of 8 and a whiteboard and pen.

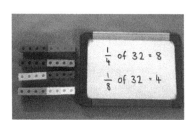

Partner 1 in each pair finds $\frac{1}{4}$ of their set of objects, while partner 2 finds $\frac{1}{8}$ using the same process as in the previous activities.

Ask if pupils can see the relationship between a quarter and an eighth of a set of objects.

Thursday

Repeat Wednesday's activity, with pupils swapping roles.

Friday

Give each pair of pupils a set of 40 objects and a whiteboard and pen.

Ask pupils to create an array with their objects, putting the objects into rows of 10. Remind pupils that 40 objects divided into 10 equal parts makes 4 in each part; equally, 40 objects divided into 4 equal parts makes 10 in each part. $\frac{1}{10}$ of 40 is 4 and $\frac{1}{4}$ of 40 is 10.

Then ask pupils to find $\frac{1}{10}$, $\frac{3}{10}$ and $\frac{5}{10}$ of 40. Pupils write their answers in the form '$\frac{3}{10}$ of 40 = 12'.

Week 3: Fractions

Recognise and use fractions as numbers

Resources: strips of paper, squared paper, cubes, bag

> **Vocabulary:** whole, part, equal parts, fraction, one whole, one half, two halves, one quarter, two quarters, three quarters, four quarters, one third, two thirds, three thirds, one tenth, two tenths … ten tenths, proportion, in every, for every, decimal, decimal fraction / point / place, numerator, denominator, equivalent, same, equal to

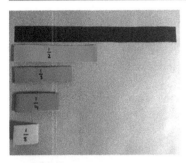

Monday

Give each pupil five strips of paper (ideally in different colours).

Model leaving one strip of paper as a whole, folding one in half, one into a quarter, one into an eighth and one into a third. Ask pupils to copy you. Remind pupils that a fraction has a numerator and a denominator. Label each strip, then compare them and discuss what the unit fractions look like.

Keep the strips for use throughout the week.

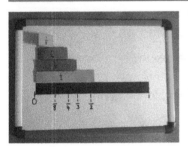

Tuesday

Give each pupil their fraction strips and a whiteboard and pen.

Pupils draw a number line the length of the whole strip of paper. They mark 0 and 1 using the whole paper strip as a guide and then mark the four fractions.

Wednesday

Give each pair of pupils a sheet of large-squared paper or a squared whiteboard.

Model drawing a fraction wall with 12 squares representing one whole. Mark $\frac{1}{2}, \frac{1}{4}, \frac{1}{3}$ and $\frac{1}{6}$ on the fraction wall. Ask pupils to copy you. Ask pupils which is greater, $\frac{2}{3}$ or $\frac{3}{4}$. Ask them to prove it using the squares.

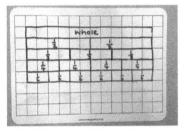

Thursday

Prepare sticks of cubes between 2 and 10 cubes long and place them in a bag; label the bag '$\frac{1}{4}$'.

Explain that in the bag are cube towers representing $\frac{1}{4}$ of different quantities. Take out a stick. Ask: 'If this is $\frac{1}{4}$, how many cubes are in the whole set of cubes?' Repeat for other sticks in the bag. Keep the bag of sticks for Friday.

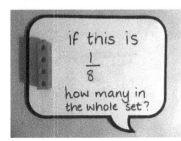

Friday

Use the sticks from Thursday but relabel the bag '$\frac{1}{8}$'.

Explain that the cube towers now represent $\frac{1}{8}$ of different quantities. Take out a stick. Ask: 'If this is $\frac{1}{8}$, how many cubes are in the whole set of cubes?' Repeat for other sticks in the bag.

Week 4: Fractions

Resources: squared paper

Vocabulary: whole, part, equal parts, fraction, one whole, one half, two halves, one quarter, two quarters, three quarters, four quarters, one third, two thirds, three thirds, one tenth, two tenths … ten tenths, proportion, in every, for every, decimal, decimal fraction / point / place, numerator, denominator, equivalent, same, equal to

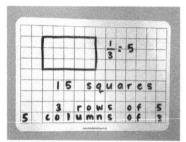

Monday

Give each pupil a sheet of large-squared paper or a squared whiteboard.

Ask pupils to use the squares to create a 5 × 3 rectangle. Ask how many squares there are in total and count them to check. Ask how many columns of 3 and how many rows of 5 there are. Ask how many squares are in $\frac{1}{3}$. Can pupils split the thirds in half to create sixths? Can they use the diagram to find the equivalent of $\frac{1}{3}$ as sixths?

Repeat for 8 × 3 and 6 × 3 rectangles.

Tuesday

Give each pupil a sheet of large-squared paper or a squared whiteboard.

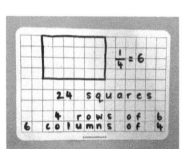

Ask pupils to use the squares to create a 6 × 4 rectangle. Ask how many squares there are in total and count them to check. Ask how many columns of 4 and how many rows of 6. Ask how many squares in $\frac{1}{4}$. Can pupils split the quarters in half to create eighths? Can they use the diagram to find the equivalent of $\frac{1}{4}$ as eighths?

Repeat for 8 × 4 and 7 × 4 rectangles.

Wednesday

Repeat Tuesday's activity but, this time, finding $\frac{3}{4}$ of the rectangles.

Thursday

Give each pupil a whiteboard and pen.

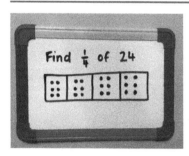

Write 'Find $\frac{1}{4}$ of 24' on the board. Model drawing a grid as shown and using dots to share 24 equally between the 4 squares to find $\frac{1}{4}$.

Repeat for other quantities which are multiples of 4. Can pupils now relate finding $\frac{1}{4}$ to finding $\frac{1}{8}$?

Friday

Give each pupil a whiteboard and pen.

Write 'Find $\frac{4}{10}$ of 40' on the board. Model drawing a ten frame as shown and using dots to share 40 equally between the 10 squares to find $\frac{1}{10}$. Then find $\frac{4}{10}$.

Repeat for other quantities which are multiples of 10. Can pupils now relate finding $\frac{1}{10}$ to finding $\frac{1}{5}$?

Week 5: Fractions

Add and subtract fractions with the same denominator within one whole

Resources: ten frames, cubes

Vocabulary: whole, part, equal parts, fraction, one whole, one half, two halves, one quarter, two quarters, three quarters, four quarters, one third, two thirds, three thirds, one tenth, two tenths … ten tenths, proportion, in every, for every, decimal, decimal fraction / point / place, numerator, denominator, equivalent, same, equal to

Monday

Give each pair of pupils a ten frame and 10 cubes.

Ask partner 1 in each pair to place a cube in each space on the frame, with the whole class counting together: 'one tenth add one tenth equals two tenths, add one tenth equals three tenths, add one tenth equals four tenths…'. When the ten frame is filled, say 'one whole' and then count back, with partner 2 removing a cube each time you count.

Tuesday

Give each pair of pupils a ten frame, 10 cubes and a whiteboard and pen.

Write on the board and say '$\frac{1}{10} + \frac{1}{10} = \boxed{}$'. Ask pupils to create this using their ten frame and then to write the answer on their whiteboard. Ask pupils to write the number sentence as shown, using a number line to reinforce or support if wished.

Repeat for other addition and subtraction questions, with pupils taking turns to answer each one.

Wednesday

Repeat Tuesday's activity, but with pupils working individually to add and subtract tenths within one whole.

Thursday

Give each pupil 6 cubes and a whiteboard and pen.

Explain that the tower of 6 cubes represents one whole. Write on the board and say '$\frac{1}{6} + \frac{1}{6} = \boxed{}$'. Ask pupils to create this using their cubes and then to write the answer and the number sentence as shown on their whiteboard.

Repeat for other addition and subtraction questions.

Friday

Give each pupil 8 cubes and a whiteboard and pen.

Write on the board and say '$\frac{2}{8} + \frac{3}{8} = \boxed{}$'. Ask pupils to create this using their cubes and then to write the answer and the number sentence in the form shown here on their whiteboard.

Repeat for other addition and subtraction questions.

Week 6: Fractions

Compare and order unit fractions, and fractions with the same denominators

Resources: strips of paper

Vocabulary: whole, part, equal parts, fraction, one whole, one half, two halves, one quarter, two quarters, three quarters, four quarters, one third, two thirds, three thirds, one tenth, two tenths … ten tenths, proportion, in every, for every, decimal, decimal fraction / point / place, numerator, denominator, equivalent, same, equal to

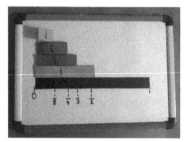

Monday

Give each pupil five strips of paper (ideally in different colours).

Model leaving one strip of paper as a whole, folding one in half, one into a quarter, one into an eighth and one into a third. Ask the pupils to copy you. Label each strip, then compare them and discuss what the unit fractions look like.

Pupils draw a number line the length of the whole strip of paper. They mark 0 and 1 using the whole paper strip as a guide and then mark the four fractions.

Ask pupils to find a fraction which is equivalent to $\frac{2}{4}$, $\frac{2}{8}$ and so on, using the strips to show you.

Keep the strips for use throughout the week.

Tuesday

Give each pupil their fraction strips and give each pair of pupils a whiteboard and pen.

Partner 1 in each pair places two strips on the whiteboard and then writes the fractions with an empty box between them. Partner 2 writes <, > or =. Partner 1 checks the answer by comparing the size of the fraction strips. Repeat.

Wednesday

Repeat Tuesday's activity, with pupils swapping roles.

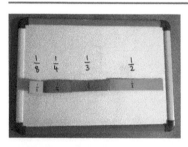

Thursday

Give each pupil their fraction strips and give each pair of pupils a whiteboard and pen.

Partner 1 in each pair challenges partner 2 to compare the sizes of the strips of paper and to write the fractions in order from least to greatest.

Repeat, with pupils swapping roles.

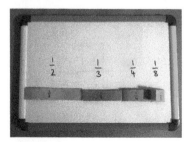

Friday

Repeat Thursday's activity but, this time, pupils must sort the fractions from greatest to least.

Vocabulary cards

Notes